AIRFIX
magazine guide 24

American Civil War Wargaming

Terence Wise

Patrick Stephens Ltd
in association with Airfix Products Ltd

First published — 1977

ISBN 0 85059 258 5

Cover design by Tim McPhee

Text set in 8 on 9pt Univers Medium by Stevenage Printing Limited, Stevenage.
Printed in Great Britain on Fineblade Cartridge 90gsm and bound by the Garden City Press, Letchworth, Herts.
Published by Patrick Stephens Limited, Bar Hill, Cambridge, CB3 8EL, in association with Airfix Products Limited, London SW18.

Don't forget these other Airfix Magazine Guides!

No 1 *Plastic Modelling*
No 2 *Aircraft Modelling*
No 3 *Military Modelling*
No 4 *Napoleonic Wargaming*
No 5 *Tank & AFV Modelling*
No 6 *RAF Fighters of World War 2*
No 7 *Warship Modelling*
No 8 *German Tanks of World War 2*
No 9 *Ancient Wargaming*
No 10 *Luftwaffe Camouflage of World War 2*
No 11 *RAF Camouflage of World War 2*
No 12 *Afrika Korps*
No 13 *The French Foreign Legion*
No 14 *American Fighters of World War 2*
No 15 *World War 2 Wargaming*
No 16 *Modelling Jet Fighters*
No 17 *British Tanks of World War 2*
No 18 *USAAF Camouflage of World War 2*
No 19 *Model Soldiers*
No 20 *8th Army in the Desert*
No 21 *Modelling Armoured Cars*
No 22 *Russian Tanks of World War 2*
No 23 *German Fighters of World War 2*

Contents

Acknowledgement

The ideas expressed in this book, whilst springing from my pen, cannot be said to have sprung in their entirety from my mind, as over a decade or more my ideas have been influenced considerably by those of others. Any person starting wargaming has also to have a base upon which to build his ideas. I should therefore like to acknowledge here the debt I owe to Don Featherstone and his early books on wargaming for my own start, and to his *Wargamer's Newsletter*, which over the years has provided a continual forum for the ideas, opinions, criticisms and debates of wargamers, many of which have altered the game as played by myself and friends. I only hope the newcomer to American Civil War wargaming will gain as much enjoyment via this book as I have gained in the past through Don Featherstone's books.

Editor's introduction

The American Civil War was, in many respects, the first 'modern' war, with breech-loading rifles being used in large numbers, the gradual emergence of the machine-gun as a weapon to be reckoned with, the end of Napoleonic battlefield tactics and the beginning of new tactics of trench warfare which were to reach their peak in 1916-18. Infantry could no longer manoeuvre in the dense columns and rigid lines of previous wars due to the devastating firepower available, so had to adopt more flexible formations. The writing was also on the wall for cavalry as a battlefield weapon for anyone with the eyes to read it, since the infantry, with their fast-firing and long-ranged rifles, could break up a charge before it made contact.

In addition to these changes, the war also saw the extensive use of railways as a means of communication for the first time, the first proper use of balloons as aerial observation platforms, and the introduction of the telegraph as a means of rapidly transmitting orders and reports.

These factors, combined with the vast scale of the war and the variety of troop types used, make it an ideal and fascinating one for wargamers, while the ready availability of Airfix 00/H0 scale infantry, cavalry and artillery sets at minimal cost make it especially suitable for the complete beginner wishing to give wargaming a try, the youngster unable to afford the more expensive metal figures, and the experienced wargamer on the look-out for an alternative period.

In this book, Terry Wise explains the basic organisation, tactics and equipment of the Union and Confederate Armies, then goes on to show how your miniature forces can be manoeuvred and made to fight in tabletop encounters. The playing rules do not cover every aspect of American Civil War wargaming, since one of the main delights of the hobby is doing your own research and formulating your own rules, but they do enable a simple, fast-moving, exciting and reasonably realistic simulation to be enacted without bogging the reader down in technicalities at this stage.

And now, on with the battle!

BRUCE QUARRIE

one

Basic principles

The American Civil War was the first truly modern war, and the first total war in the modern sense. In fact it is often credited with a long list of firsts, the main ones being the first appearance on the battlefield of large quantities of rifled small arms and artillery, and repeating rifles; the first use of railroads as a major means of transporting men and supplies; the first extensive use of trench warfare; and first widespread use of telegraphic communications, resulting in the rapid dissemination of information direct from the battlefield. These 'firsts' revolutionised tactics and strategy and were to have far-reaching effects.

There is another 'first'; the American Civil War used to be the war with which many people began wargaming — mainly due to the fact that in the early 1960s Airfix produced a range of cheap American Civil War figures. In more recent years a vast range of Napoleonic figures has become available, both in plastic and in metal, and the colour and dash of Napoleonic wars has ruled supreme for some years. However, I am sure a revival of interest in the American Civil War is on its way and many Napoleonic buffs will soon be painting new armies, or dusting off old ones to return them to American Civil War battlefields!

Although the American Civil War is a 'modern' war, with trenches, sieges and victory by attrition, at its outbreak in 1861 weapons, strategy and tactics were still of the horse-and-musket era. The men marched to war in uniforms as colourful and fancy as any of the Napoleonic era, armed with smooth-bore, muzzle-loading muskets and artillery, and in some cases still with flintlocks. They formed up in tight ranks and closed with the enemy for a musketry duel at short range, followed by a swift bayonet charge if successful.

This was the situation in the first half of the war but gradually the introduction and perfection of new weapons brought about a change to more stagnant warfare, with one side taking up a defensive position and the other side unable to capture it, even with odds of three to one, without suffering casualties so severe as to make any victory Pyrrhic. Gettysburg, July 1-3

Rebels make a dashing but foolhardy frontal assault against entrenched Federal troops during a wargame set in 1863.

A wargame set in 1861 with CSA forces attempting to seize control of a turnpike. Note the extensive cavalry melée, backed by solid blocks of troops, and the open battlefield.

1863, is often quoted as the turning point.

It is this change during the course of the war which I feel has made, and will make again, the Civil War so popular with wargamers. The Napoleonic player can enter a new yet familiar field with the first half of the war, and find himself led into the future; the 'modern' player can recreate the second half of the war with its devastating firepower and find himself investigating how it developed from the earlier fighting; and the newcomer can plunge in feet first with the knowledge that once he has mastered the American Civil War he will be able to spread in both directions with a fair grasp of the weapons, organisation and tactics of both eras.

The aim of this first chapter therefore is to provide those newcomers with sufficient basic information to enable them to recreate the Civil War on a wargames table and understand the main principles of the rules. So let us first see precisely what is needed to set up a game.

First, two opposing 'armies.' Airfix produce plastic figures which enable Federal and Confederate armies to be built up for a small outlay: details of these and of metal figures available are at the end of the book. Airfix figures are 00/H0 scale, which means they stand between 20 and 25 mm high. Most metal figures are 25 mm, though some from the USA are 20 mm. Unfortunately it is not really possible to mix metal and plastic figures because of the scale variations, for although men vary in height, wargamers prefer to see all their men the same height. Personally I regard this aesthetic approach as the first indication that wargaming is for fun and not an attempt at achieving absolute realism in miniature: no real army has all its men of identical height! The model figures are painted in the uniforms of the period—this research and painting can be a rewarding aspect of the hobby by itself—and organised into battalions, squadrons and batteries.

The next step is a battlefield, usually the dining table with a smooth green cloth over it, or a plywood, chipboard or framed hardboard playing surface placed on the table to create a larger playing area. The size of the playing area depends entirely on the space

American Civil War Wargaming

available and may be anything from 5 × 3 feet to 9 × 6 feet or more. No playing area is ever big enough, as you will find your armies always expand to fill it, on the basis of 'If I'd had just one more battalion I would have won!'.

Since no battlefield is like a billiard table, terrain features are added. Trees, stone walls, buildings, bridges and even river sections may all be purchased in model shops. Roads can be cut from card or vinyl floor covering; fields can be made from corrugated cardboard or hardboard with walls or fencing round the edges. Hills are usually represented by contours, that is rounded ceiling tiles or chipboard stacked on top of each other to the required height. The model figures are set out and manoeuvred on this model terrain, their movement, firing, state of morale and the casualties suffered all controlled by a set of rules.

Unlike other major wargaming periods, such as the Napoleonic Wars and World War 2, there are (at the time of writing) no universal rules for Civil War wargaming and, more than in other eras, each club or group of players tends to stick to the rules they have drawn up for themselves; rules which are constantly changing as fresh information is discovered, new ideas tried out, old ideas dropped because they failed to produce the required results. And periodically, as the rules

A reconstruction of the ACW battle of Murfreesboro, fought as a wargame, with the Federal line forced back and Confederate attacks going in.

become too unwieldy, people revert to their original simple rules of two or three pages in order to just gain the maximum enjoyment from a game.

You may have gathered by now that wargaming is a highly individualistic hobby, and there are really no rules except those you make yourself, or those of others which you traditionally

A selection of the figures available in the Airfix US infantry, CSA infantry, US cavalry and ACW artillery sets.

Some of the figures from the Airfix Cowboys and Wagon Train sets which can be used for ACW armies.

A wargame terrain set up with Bellona stream lengths, bridges and walls; Mini-tanks and Merit trees; home-made roads (cardboard), fields (hardboard), and bushes and hedges (sponge).

American Civil War Wargaming

A basic layout with the boundaries of two dense woods marked in chalk, the remainder of the table unadorned. Lack of cover and room to manoeuvre made this a game of attrition.

Hills can be made from plaster, which is more realistic than other materials, as may be seen here, but in practice such hills have a limited use because of lack of space for figures and the half inch chipboard used in layers to represent contours is far more effective.

Basic principles

The quiet before the storm: a wargames table set up with elaborate scenery. The background is made from railway scenery paper on hardboard.

accept when playing against them in their home or club, and perhaps use as a basis upon which to create your own rules. It follows therefore that the rules outlined in this book will be *my* rules! You are welcome to accept them, reject them, discard parts, enlarge on other parts; this is how I got mine. I make no excuses for being a wargamer who merely enjoys the game, but I have tried to include sufficient information in the following chapters for anyone wishing to formulate more complex rules to do so. My own preferences are included in the hope that they may give enjoyment, or at least inspiration for more ideas, to others.

Additional complications in the case of Civil War rules are the two phases of the war with two different types of warfare, and the vastness of the USA, which meant several theatres of operations, each with its own characteristics. It is not really possible to cover such wide variations in one set of rules and a wise move at this stage is for the player to ask himself which half of the war he would like to concentrate on first

(always remembering you can develop the rules to suit the second period) and whether he would prefer the close-knit east coast fighting or the more widely spread western theatre. The information supplied in the next three chapters should give some help, but most beginners would be advised to start with 1861 in the east. The rules listed later are intended for the actual playing of a game and for space reasons (and ease of playing) they are rather general in scope, although they lean towards 1861-3 in the east.

Figure, ground and time scales

Although there are no universal rules for our period, there are basic principles for all rules, and the most important of these are figure, ground and time scales, for upon the understanding of these depends the success of any set of rules. I will deal with them in some depth, for once you grasp them you are well on the way to a lifetime of enjoyment.

American Civil War Wargaming

Another plaster hill, this time with a firing step carved out.

Civil War regiments averaged 500 men and it is obviously not possible or even desirable to have one model figure per man. The most popular method of scaling down such numbers is one model figure equals 33 men: with this scale a regiment can be represented on the table by 15 model figures. Officers are not included in this representation and usually count as one man for a

Federal troops battle with Indians fighting for the CSA in the Indian Territory.

A struggle for a New England town, with both players feeding troops into the fight from reserves 'off the table.' Skirmishers in the wood at top left are Federal, this player being nearest the camera.

Another triumph for the CSA! The end of a game named Bitter Creek, showing Confederate regiments in line, two deep. In the centre a depleted regiment illustrates that casualties may be removed despite the use of bases to speed up moves.

American Civil War Wargaming

Model guns should also occupy the ground area taken up by an entire battery; 82 yards frontage for a 6-gun battery, 55 yards for a 4-gun battery. This same-size base illustrates the type of base for a 4-gun battery. The dotted lines indicate the angle of fire permitted without changing front.

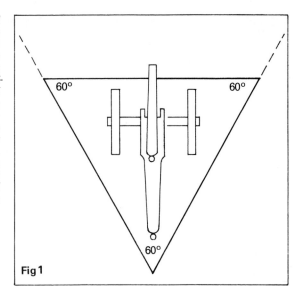

Fig 1

number less than 33. However, the organisation of Civil War regiments is such that this ratio does not fit conveniently and I prefer a ratio of one model figure equals 25 men. An artillery battery is usually scaled down to one model gun and crew. More details of these ratios are given in the next chapter.

In the same way that we scale down the figures, so we scale down the terrain over which they move. Battles of this period were fought over fronts varying from one to six miles and if we take the 25 mm figure's 4 mm = 1 foot scale as our ground scale, we would have a tiny combat area represented by the wargames table. Therefore, it is customary to ignore the figure height and settle on a ground scale of 1 mm = 1 yard, giving a good sized battlefield of almost one and a half miles by just under a mile on an 8 × 5 foot table.

This means we have an anomaly between ground scale and vertical scale, of which more in a moment, but this is unimportant—except aesthetically—for only the ground area occupied by a model figure's base is critical. The average base area for model soldiers is about 10 mm square, so a model figure is occupying an area of ten square

yards—room for 100 men in close order. However, Civil War regiments went into action in open order, two ranks deep, so a regiment of 500 men would have a depth of only ten yards but a frontage of 250 yards at one yard per man. To scale this is 250 × 10 mm and therefore in open order (if 1 figure = 25 men) our model soldiers need a base frontage of 12.5 mm to obtain the correct ground coverage. Unfortunately it is not always possible to obtain the correct ground scale for depth, as in this case where we have a required depth of 10 mm but two ranks of figures occupying a depth of 20 mm. This is unavoidable and can only be kept to a minimum by making the movement bases, on which wargames figures are usually glued, as narrow as the figures' bases allow.

So we have a figure base requirement of 12.5 × 10 mm. To speed up movement of figures on the table it is usual to stick several figures on one base of a correspondingly larger size, using thick card or floor tiling. Because two figures represent a company in our organisation, it is best to have no more than two figures to a base, 25 mm long. To allow individual figures to be removed as casualties, and to operate

Early stage of the Bitter Creek game. Here a cluster of three buildings represents a village, and beyond them a circle of trees represents a small wood.

some companies in extended order as skirmishers, it is recommended the 20 figures of a regiment be grouped in seven sets of two and six single figures.

Cavalry operated in fairly loose order and a base frontage of 25 mm best indicates this formation. Depth of base will be governed by the model. I

A is the actual line of vision because at our ground scale the figures would be 25 yards high! B is the scale line of vision taken from a ground scale level of five feet or 1mm, which is the top of the figure's base.

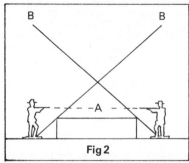

Fig 2

prefer to mount each cavalry figure on an individual base for convenience in mêlées.

To return now to vertical scale. For visual reasons this is usually in proportion to the figure scale of 4 mm = 1 foot. Thus buildings, trees, fences etc are all in proportion to the figures, *but* two or three houses will represent a village—we have to consider the ground space being occupied—and a small clump of trees represents a wood, usually with the centre left bare to make movement of figures easier. The result is reasonable realism in scale but a layout which is also pleasing to the eye.

The one really important item in vertical scale is high ground. Usually represented by layers of polystyrene or chipboard rarely higher than 15 mm, it is accepted that each layer represents a contour of 40 feet. However, because a 25 mm figure can 'see' over such a contour does not mean units facing each other across it may exchange fire. The same problem does not occur with buildings and trees, which are physically higher than the figures. For convenience these too are normally assumed to be an average 40 feet high.

Lastly there is the time scale. Because the distance a unit may move at any one time, or the number of times a man can fire his weapon, is governed by the length of time given for such movement or firing, there must be a direct link between time scale and movement on the table if a game is to achieve any degree of realism. It is generally accepted that 2 ½ minutes is a good period of time, and each game move is presumed to have represented the actions of the figures in 2 ½ minutes. Therefore a man capable of marching 110 yards a minute would cover 275 mm in a game move. It is usually found a time scale of more than 2 ½ minutes allows units to move too far in one game move and the moves have then to be split to see where units were at particular times.

One final word on scale — casualties taken from the table. Because each figure represents 25 men, casualties are often reckoned in real numbers and a model figure removed every time a total of 25 is reached. The firing tables in my rules work on percentages so this casualty problem does not arise but if you use another method of inflicting casualties it is vital you remember one figure represents 25 men if you wish to have a realistic game.

two

Organisation

The total strength of the Confederate Army was never more than just over half that of the North — most of the time it was below that figure — and at the outbreak of war it consisted entirely of inexperienced volunteers and poorly trained States' militia. The Union Army had also to rely almost entirely on militia and volunteer companies because nearly all its 16,000 regulars were tied up in frontier posts and sea coast forts: at Bull Run only 2,000 of McDowell's 38,000 troops were regulars. These facts are important when organising Civil War armies, for the CSA should have a smaller army than the North and a distinction must be made between experienced troops and recruits.

The two sides also set about organising these inexperienced armies in different ways, yet with the formations almost identical because the

A Wisconsin regiment of 20 figures, with two companies advanced as skirmishers and another two companies in reserve.

senior officers had all trained at West Point. In the CSA ex-regular officers were used to train new regiments and of 304 officers who had been on active service in the Federal Army before the war, 148 reached general officer rank. This gave the Confederate Army good generals and had a marked effect on their troops' fighting value — reflected in better morale factors for CSA generals in the rules. In the Union Army the 780 regular officers remaining loyal were mostly kept in their regiments. The result was poor leadership for the first two years, with new regiments led into battle by inexperienced officers and suffering the inevitable initial heavy losses. So it may be seen that while the Confederate Army should be smaller, its fighting value may be kept on a par with that of the Union Army.

Both armies were divided into the three main arms of infantry, cavalry and artillery, and we will now deal with the organisation of each arm in detail.

Union and Confederate infantry

The official strength of a Union infantry regiment was between 869 minimum and 1,049 maximum, divided into ten companies and with 35 officers. Confederate regiments followed the same organisation but for a brief period at the start of the war also had some legions, ie a mixed force of infantry, cavalry and artillery, acting as a single unit.

These were theoretical strengths and even in the early days of the war, when enthusiasm was at its height, even volunteer companies could not get all the men they needed. Often there were between 700-800 men to a regiment when it reached the front line and sickness, non-combatant duties and the first action usually brought the number down to an average of perhaps 500 — our 20-figure wargames unit at 1 figure = 25 men.

The two sides adopted different methods for dealing with recruits. In the CSA recruits were fed into existing regiments: this helped to keep alive an *esprit de corps* and had a good effect on

A Confederate regiment of 20 men and two officers 'defending the rum ration!'

morale. In the US Army regiments were allowed to fall to between 150-200 men, when they were disbanded and the survivors distributed amongst new regiments. This meant loss of regimental pride and of experienced units. It also meant that on average the CSA regiments tended to be stronger than many Union ones and, having a hard core of experienced men were, at least until mid-1863, more effective as a fighting force.

In 1863 the War Department tried to adopt a programme of sending recruits to existing regiments but by then there were more than 1,000 regiments and to have filled them all up would have created an army far beyond the needs of the Union. Therefore, even late in the war, new Union regiments continued to go into battle under inexperienced officers and were allowed to become smaller through losses, their lack of numbers being compensated by the experience so expensively gained.

To reflect these differences in wargames the US Army regiment is averaged at 375 men (15 figures with 1½ to a company), the CSA regiment at 500, or 20 figures with two per company.

There was a single exception in the Union Army to this practice — the Wisconsin regiments, where old regiments were 'topped up' as in the CSA. Sherman wrote in his memoirs 'we estimate a Wisconsin regiment equal to a normal brigade. I believe that 500 new men added to an old and experienced regiment were more valuable than 1,000 men in the form of a new regiment.' Wisconsin regiments should therefore be based on the CSA unit of 20 figures.

In May 1861 the US Government called for nine new regular regiments, to be two or more battalions with eight companies to each battalion. These were the 11-19th Regiments. Full strength was never achieved and of those regular regiments at Gettysburg none had more than eight companies present. These regular regiments can be 16 figures in eight companies on the wargames table, equal to a strength of 400 men.

A few regiments were raised for special duties, or may be classed separately. One such type was the Zouave regiment which, even after the fancy uniforms had disappeared from other regiments, clung to its baggy

Organisation

A reconstruction of the battle of Cedar Run with the CSA forces in retreat after the arrival of Rickett's 2nd Division halfway through the game. Note skirmishers covering a withdrawal on the left flank and a line of retreating skirmishers in the centre who have been covering re-alignment of CSA forces there.

trousers and distinctive cap—at least until mid-1863. There was a certain *esprit de corps* in these regiments, a feeling of being an élite, and this was

Confused mêlée between surprised Union cavalry and mounted and dismounted CSA Indians in the western theatre.

furthered by the issue of the M1862 Remington or Zouave rifle musket to those Zouave regiments which survived the first year of the war. (The musket had a faster rate of fire and longer effective range than the weapons of most other regiments.)

Another special corps was the Sharpshooters, raised by the Union in the summer of 1861. Eight companies formed the 1st and 2nd Regiments but the men fought mainly in small groups, sniping at officers and guns' crews. The two regiments served with the Army of the Potomac and were eventually grouped as a single regiment in 1864.

The CSA also had snipers, though not on any organised lines, and there were probably no more than about 100 at any one time.

Negro troops were recruited by the North as early as April 1862 and the first regiment was organised in July that year as the 1st South Carolina Volunteers. The 54th and 55th Massachusetts Regiments followed and finally there were 120 Negro regiments.

There were three million slaves in the South (a third of the population) but this vast resource remained untapped, and the organising of slaves as soldiers

American Civil War Wargaming

was not authorised until after February 1865, by which time it was too late. One unit of some 125 Negroes serving the Confederacy was reported at Petersburg.

Similarly, in the autumn and winter of 1861 the CSA organised three regiments of Indians from the Choctaw, Chickasaw, Cherokee, Cree and Seminole tribes for service in the Indian Territory (modern Oklahoma and part of northern Texas.) They first saw action in Arkansas at Pea Ridge, March 1862. Parts of four regiments and a battalion were subsequently raised in the same tribes but the total number of men does not seem to have been above 3,500 at any time. A small wargames unit of about ten figures could be used in the western theatre.

The North raised three regiments of Cherokee, Cree and Seminole Indians in July 1862, again for the Indian Territory. These regiments are believed to have been each 1,000 strong. The 53rd New York had a company of Tuscarora Indians, and another New York regiment had a company of Senecas. The Indian regiments of both sides were divided into companies as usual and were commanded by white officers but with Indian company commanders and below.

From the above it can be seen that varied and colourful armies may be organised for the pre-Gettysburg era. A Federal army could have one each Negro, Zouave, Wisconsin and Regular regiments, a maximum of eight Sharpshooter companies (two figures per company) if the Army of the Potomac, Indians if in the west, and a variety of volunteer and militia regiments. The CSA forces could have a Zouave regiment, plus Indians if in the west; other regiments would be militia or volunteers.

Higher organisation

Regiments were grouped in brigades of four (often from the same state in the CSA Army), but the number could vary from three to six and to be precise there was really no permanent formation higher than the regiment; brigades and above being more akin to the modern battle group, assembled from the units

A brigade of Confederate infantry advances in column, supported by artillery and cavalry, with skirmishers protecting one flank and the fourth regiment of the brigade well in advance with half its strength deployed in extended order.

Lincoln with the Federal staff after their victory at Antietam.

available to the best combination of arms for the task ahead.

A division could be from two to five brigades, a corps from two to four divisions. Armies were made up of any number of corps. The brigades and divisions of the Confederate Army tended to have more regiments than the corresponding US Army formations.

Union cavalry

Soon after the outbreak of war the US regular cavalry regiments were increased from five to six and numbered 1-6, each with six squadrons of two companies each. At first each company was 100 men, a captain and three lieutenants, but in 1863 the regulation strength was changed to 82-100 men, a captain and two lieutenants. At the same time the squadron was dropped and the battalion of four companies introduced, mainly for detachments from the main body. In practice strengths were usually at the bottom of this scale so we take 75 men for our wargame company, or three figures, with 12 figures for a battalion. 36 figures could be used for a regiment, but I prefer to

have a greater variety of battalions than a few large regiments.

In the first two years of the war the cavalry was sometimes attached to infantry divisions but early in 1863 the cavalry of the Army of the Potomac was organised as a separate entity right up to corps level and the other armies subsequently followed suit. A brigade was four to six regiments, a divison two to three brigades, and a corps two to three divisions.

There were seven Negro cavalry regiments out of a total of 272 cavalry regiments raised, so a Negro battalion could be included in your Union Army.

Confederate cavalry

The Confederate squadron had a theoretical strength of 60-80 men, a captain and three lieutenants per squadron, and ten squadrons to a regiment. In fact regiments rarely reached these strengths and a realistic figure would be 50 men per squadron, giving a wargames squadron of two figures and a regiment of 20 figures.

A brigade could be two to six regiments, a division any number up to six brigades. The CSA does not appear

American Civil War Wargaming

to have organised corps of cavalry as such.

A large number of Indian regiments served the CSA from early 1862 and in 1864 were organised into the Indian Cavalry Divison of two brigades, supported by Texan artillery and cavalry. They were restricted to the western theatre.

Two other aspects of the cavalry which should be taken into consideration are skill and horse supply. In the first half of the war the Confederacy had born cavalrymen and a good supply of horses: the Union had neither. In the second half of the war the Union had both, but the Confederacy was short on horses and many veteran cavalrymen were forced to become infantry. To allow for this in wargaming it is best to give the Union Army a lower proportion of cavalry until 1863, the CSA a lower proportion in 1864-5. The proportion of cavalry to infantry should never exceed 1:6 for either side.

Artillery

Union artillery batteries normally had six guns, but four or eight were used occasionally. We take six as the average and place the model gun on a triangular base, as shown in chapter one, with each side measuring 82 mm. 12pdr batteries usually had four 12pdrs and two 24pdr howitzers: 6pdr batteries had four 6pdrs and two 12pdr howitzers. Later in the war batteries usually had all one type and calibre guns.

Crews were about ten men per gun maximum. As there were spare crews and other personnel, six figures per model gun is a reasonable balance.

Most wargamers do not bother with limbers, certainly not caissons and baggage wagons, but it should be pointed out that each gun had a limber, a caisson with ammunition, and the battery as a whole had extra caissons, a baggage wagon to each gun, and a field forge. This clutter of vehicles often

Cavalry mêlée in the reconstruction of Murfreesboro. In the background Beckenridge is about to make his fatal charge. Federal troops are hurrying across the river to reinforce that flank.

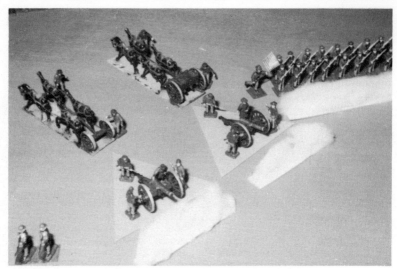

Two Union batteries in action, supported by infantry, showing the battery front represented by the artillery templates and the permitted angle of fire.

caused problems on a battlefield, where the gun and caisson horse teams were drawn up behind the guns to a depth of 47 yards. They should therefore be represented by *at least* a two-horse limber and if possible a four-horse caisson as well. Such models will cause loss of manoeuvre—as indeed they did on the battlefield—but they should not be aligned deliberately as 'defences' nor should cavalry attempt to vault over them!

Fierce fighting on a flank during a wargame. Two brigades of CSA infantry are advancing in column from the centre of the table. Note particularly the cluster of limbers behind the massed guns at top left; quite a barrier to movement if those columns should later have to move rearwards and to the left.

American Civil War Wargaming

Murfreesboro as a wargame at about 10 am on the second day as the Confederate cavalry cut the railroad at the rear of the Union forces. In the real battle many Federal troops had begun to retreat by 11 am because they were out of ammunition as a result of this manoeuvre.

Whilst on the subject of wagons, etc, it is worthwhile mentioning that armies were dependent on their supply trains: such trains should always be represented at the rear of your army on the wargames table. This will force you to tie up a proportion of your army to defend the train, and provides a secondary objective in a game, particularly for roaming cavalry. This is as it should be: armies did not march on empty stomachs, nor could they fire their weapons forever without replenishment.

Confederate batteries advancing at top speed on a flank.

Union artillery was normally allotted at the rate of four batteries per infantry division, with often half this force withdrawn to form a corps reserve. Each army also had a reserve of light and heavy batteries. From Gettysburg on, however, the Army of the Potomac usually concentrated its artillery at corps level, with nine batteries per corps.

There were 12 heavy artillery and one light (horse) artillery regiments of Negro troops in the Union Army, so a battery of Negro artillery could be included in your army. However, Union heavy artillery regiments (both Negro and white) were often used as infantry in 1862-3, when the heavy guns were not needed so badly as the men. These regiments were often full strength and had 12 companies; a tempting unit for a wargamer! There were also Regular Army artillery regiments as opposed to volunteers, though the volunteers seem to have been equal to the regulars after the first few battles.

Confederate batteries were most often of four guns, sometimes six and occasionally eight. In this case we take four as the average and place our model gun on a triangle of card with 55 mm sides. A crew of four is used, and these two factors reflect the Confederacy's weakness in artillery.

Four CSA batteries made a battalion and there was normally a battalion to an infantry division, with further battalions as corps reserve.

The proportion of guns in a model army should be no more than one model gun per 60 figures for the Union; one model gun per 70 figures for the Confederacy.

three

Tactics

To achieve a fair degree of realism, wargamers should follow the tactics of their period as closely as possible, ignoring hindsight and modern developments, and wargame rules are where possible shaped in such a way as to make it worthwhile for players to conform to the tactics of their era and punitive to attempt tactics which are not applicable to that era. The next step in metamorphosis for real armies to wargames ones is therefore a knowledge and understanding of the tactics of the era: how the units organised according to the information in the previous chapter, are deployed on the table, their formations, and methods of attack and defence.

Infantry tactics

Long marches were made by a regiment in column of fours (four abreast) or sometimes in column of companies. On the wargames table a regiment has to be in column of companies, ie two model figures abreast. This formation should be restricted to the approach march and should be changed to line of two ranks as soon as possible. To encourage this, troops fired on in column suffer much heavier casualties than troops in line.

On the field of battle regiments normally fought in line with skirmishers forward to shield the main force and weaken that point of the enemy line to be attacked. Sometimes up to half a regiment might be used as skirmishers and, if a divisional attack was being launched, whole regiments might be committed in this role.

Attacks were launched in waves of regiments in double ranks, preceded by the skirmishers and with 250-300 yards between waves. This spacing permitted the following regiments to turn at right

A Union brigade with artillery and cavalry support attacks a fortified position in the Confederate line. The attacking regiments are deployed in line, with skirmishers on the flank. Reserves are advancing in column in support.

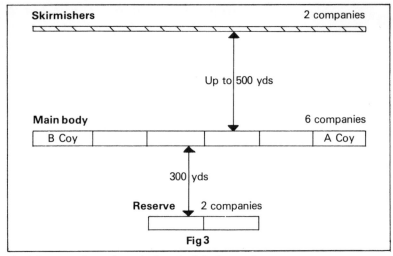

Skirmishers		2 companies

Up to 500 yds

Main body		6 companies
B Coy		A Coy

300 yds

Reserve 2 companies

Fig 3

An infantry regiment in typical tactical deployment.

angles to their route if the flanks were threatened and to support the first line without receiving fire directed at the leading regiments. The heavy casualties inflicted by canister at close range encourages players to keep their regiments in this open formation!

The aim of an assault was to get close enough for a crippling exchange of musketry. Very few bayonet charges were made. General J.B. Gordon of the Confederate Army wrote in his memoirs: 'I may say that very few bayonets of any kind were actually used

Union cavalry making its way through a gully in column, and ripe for ambush.

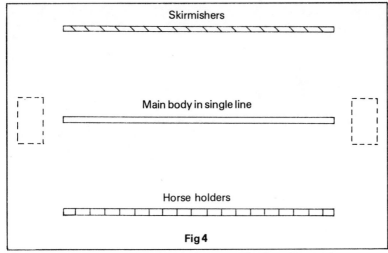

Skirmishers

Main body in single line

Horse holders

Fig 4

A cavalry regiment in typical tactical deployment. The dotted rectangles represent mounted companies often kept on the flanks.

in battle, so far as my observation extended. The one line or the other usually gave way under the galling fire of small arms, grape, and canister, before the bayonet could be brought into requisition. The bristling points and the glitter of the bayonets were fearful to look upon as they were levelled in front of a charging line; but they were rarely reddened with blood.' Of some 250,000 Union soldiers wounded in the war, only 922 had sword or bayonet wounds. In fact, because of the deadly performance of the weapons in use after the first few months, charges often degenerated into short rushes by small groups of men, using all the cover they could find. Eventually the men simply refused to advance in the open against the firepower now available to both sides, and in the second half of the war there was a great deal of trench warfare.

Cavalry tactics

Cavalry manoeuvred in column of fours for maximum flexibility; there is no choice but to use two abreast on the wargames table for this. The fighting formation was a double line until 1862 but during that year both

sides changed to a single line. Again, cavalry in column suffers more casualties than cavalry in line, and this encourages players to use mostly the line formation.

It should be remembered that successful charges were generally only made over clear ground; any form of obstacle in the terrain could disrupt a charge, possibly injure men and horses, and certainly enable a quick-witted enemy to catch the attackers at a disadvantage. This factor is also reflected in the rules by introducing penalties for those charging over rough ground.

The cavalry did not fight alongside the infantry because it could no longer charge the infantry due to their increased firepower, but it fought a variety of actions with the enemy cavalry, including sabre and revolver mêlées, dismounted actions with firearms, and combinations of both these.

Early in the war the Union cavalry, being at that time inferior to that of the South, adopted the role of dragoons; exploiting their mobility to seize advanced positions until the arrival of the infantry, or to cover gaps in the battle line, or to cover retreats. Unfortunately cavalry carbines did not stand up to

A CSA battery engaging Federal infantry with shrapnel. Canister will be used if the attack is pressed home.

prolonged firing, nor had they the range of infantry weapons. This meant cavalry could not put up a prolonged resistance to infantry attack, and the dragoon role was therefore limited, even after 1864, when the US cavalry began to be armed with the devastating repeating carbines.

The Confederate cavalry became famous for its dashing raids during this same period.

From 1863 the scene began to change. Cavalry mêlées were still fought, but now they tended to use carbines and revolvers instead of sabres, and troopers often had two or four revolvers, two at the waist and two in saddle holsters. Confederate cavalry in the western theatre favoured shot-guns, fired at close range at full gallop, then used their revolvers in the mêlée.

The greatest cavalry battle of the war, Brandy Station, June 1863, involved almost 20,000 cavalry for more than 12 hours, and at the height of the battle charges and counter-charges were made continuously for almost three hours. After Gettysburg Imboden's cavalry guarded Lee's re-treat in a series of whirlwind mêlées

with Buford's Federal cavalry, but the pattern for all these mêlées was the same — hit fast, do as much damage as possible, and break away quickly. This hit-and-run type mêlée is the only one possible with the rules set out in this book.

Artillery tactics

When an infantry attack was launched the batteries armed with rifled guns would open fire with shot at their longer range, next the smooth-bores would join in with shot as the enemy came closer, then the artillery would switch to spherical case, and finally discharge canister at close range. Canister was so effective against infantry attacks that the smooth-bore remained popular in both armies throughout the war: in McClellan's Peninsula Campaign his chief of artillery specified that two thirds of the Army's guns should be smooth-bores. The 12pdr smooth-bore Napoleon was the most frequently used gun of the war.

Counter-battery fire could be carried out but the main value of artillery was in defence, breaking infantry attacks before they could get into close musket

The cavalry column ambushed by Confederate Indians and making a stand on a ridge. Note the use of infantry figures to represent dismounted troopers.

range. It was therefore normally employed against the infantry, and at Gettysburg the Union batteries were actually ordered to cease counter-battery fire so as to conserve ammunition and allow the guns to cool for the infantry attack which would follow the artillery duel.

Once the bulk of the infantry was armed with rifles the artillery was forced even more into a purely defensive role for it could no longer advance in close support of infantry attacks, the enemy's infantry being able to pick off the gunners before the guns could get within canister range.

Artillery was also threatened by sharpshooters and marauding cavalry, the latter falling unexpectedly on the batteries whenever the opportunity presented itself.

Indians

Indians were mostly allowed to fight in their own way—achieve surprise, hit the weakest point, and withdraw quickly when sufficient damage was done. The Confederacy's Indian cavalry fought on foot, leaving their tethered ponies at the rear, and on horseback. As mounted troops they were particularly effective at harassing Federal lines of communication, but as infantry they lacked the long-range firepower of the white troops. Indian tactics are encouraged in the rules by keeping them in extended order and giving a bonus in mêlées.

four

*Boys, this will be short but desperate

We now know how Civil War armies were organised, the tactics they used and how to apply the rules of scale to reproduce these armies on a wargames table. Now we must study how our wargame armies will move and fight on the table top. As in the previous chapters, we take the real life data, or as close to it as we can get, and scale it down.

Movement

The US Drill Manual for our period states infantry columns could move at 70 yards per minute in common time, 85 yards quick time, and 110 yards double quick time. The latter would have been used mainly for advancing reserves in column to a point of danger and for charges, so we restrict this speed to columns and charging infantry. Infantry in line would normally advance at common time, unless charging to close with the enemy. This was a rare occurrence in the Civil War and mostly the infantry advanced more slowly, stopping periodically to fire. Because it is more difficult to maintain formation in line than in column, common time for line is estimated at 60 yards a minute. Skirmishers would move quickly without regard to formation, so we can use quick time for them at all times, double quick time if running back to the main body to escape attack.

Multiplying yards per minute by the 2½ minute game move we get: common time, $60 \times 2\frac{1}{2} = 150$ mm per game move; quick time, $85 \times 2\frac{1}{2} =$

*Confederate General Strahl at the battle of Franklin.

212, say 225 mm per game move: double quick time, $110 \times 2\frac{1}{2} = 275$ mm per game move.

Charges can only be made for the final move which results in a mêlée. Double quick time should be restricted to two moves, followed by one move at common time.

Cavalry had six speeds: walk, 4mph; slow trot, 6mph; manoeuvring trot, 8mph; alternate trot and walk, 5mph; manoeuvring gallop, 12mph; full gallop, 16mph. For the wargame these can be reduced to cavalry in line moving at the walk of 120 yards per minute ($\times 2\frac{1}{2}$ minutes $= 300$ mm per move) with a bonus when making a charge. The full gallop was usually employed only for the last 50 yards of a charge so in a 2½-minute move we have $2 \times 120 + \frac{1}{2}$ minute at 480 yards, giving a total charge move of 480 yards, say 475 mm. Column would move slightly faster than line, so we take the alternate trot and walk of 5mph, or 146 yards per minute, $\times 2\frac{1}{2} = 365$ yards, say 375 mm.

Field artillery rode their guns and caissons into battle, while in the light batteries every man had his own mount. This meant light batteries could move at a similar rate to cavalry in column, 375 mm. Field artillery would have been slightly slower, the cavalry line speed of 300 mm, and heavy batteries slightly slower still, say 250 mm.

Infantry firing

The US Army regulations of 1860 specified a trained soldier should be able to fire three aimed shots per minute—this being with the muzzle-loading smooth-bore musket in general use at that time. As our game move equals 2½ minutes, this means a trained soldier armed with such a musket should be able to fire, say, seven times per move. However, not all soldiers were trained when they went into battle and sometimes even trained men behaved erratically when under fire for the first time. After Gettysburg, for example, of some 37,000 muskets salvaged from the battlefield, 18,000 had been loaded twice (or more) without being fired, showing that in the frenzy of battle many men had forgotten their drill. From these figures it has been estimated that the firing of 35 per cent of the troops engaged in the battle was ineffective! Another common mistake was to forget to remove the ramrod after reloading, thereby discharging it at the enemy and preventing further reloading, at least for some time.

It can be seen, therefore, that only veteran troops were likely to fire at top speed; others would perhaps average only half the veterans' speed. To achieve this distinction between various regiments, my units are divided into veterans, experienced troops or recruits. The percentage of hits likely to be scored in a set time — such as our game move — are set out in table form in the rules chapter and it is necessary only to consult these tables to determine how many casualties are inflicted within each game move. The tables take into consideration known percentages of hits by volley firing at various ranges, and allow for varying percentages of such hits depending on the quality of the troops, the number of men firing, and the type of weapon being used. The performances of the real weapons are listed overleaf for comparison with the tables.

Infantry fire-fights were most often decided by the morale of a unit and casualties were roughly equal when the approximate same number of men and type of weapon were involved. The firing tables, with their standard rates of casualties, emphasise the importance of morale and at the same time eradicate the unpredictability of results obtained with dice and cards. Because the tables can be easily consulted (I paste them on postcards) the game is also speeded up, while the inclusion of many ranges, instead of the customary three or four, means there are fewer 'hard lines' such as where firing at a range of 150 mm kills two figures, but at 151 mm kills only one.

At the outbreak of war both sides

An ACW game seen from the Confederate side, with, on the left Federals attacking a farm, on the right a Confederate brigade launching an attack against another farm. Both sides have avoided an advance across the open centre, the Federals having advanced from the cover of a wood, the Confederates from behind a hill.

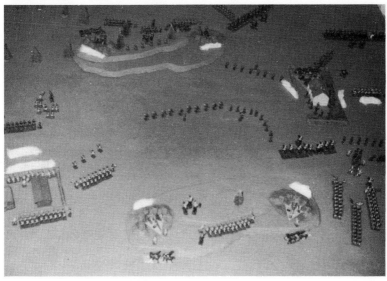

Boys, this will be short but desperate

Infantry weapons

Weapon	Rounds per minute	Range in yards Max	Range in yards Effec- tive	Range in yards Battle	Remarks
US Percussion M1842 muzzle-loading smooth-bore	2-3	300	150	75	Standard weapon in 1st months of the war.
Springfield M1861 muzzle-loading rifle	3	1,000	500	250	Most common weapon on both sides.
Enfield muzzle-loading rifle*	2-3	1,100	500	300	2nd most common, more accurate than Springfield
Remington (Zouave) M1862 muzzle-loading rifle	3	1,200	600	350	3rd most popular, more accurate than Springfield.
Sharps breech-loading rifle M1863	8	1,800	600	350	Issued to US sharp-shooters 1863.
Spencer breech-loading rifle	16	1,800	600	350	Union troops only. Those captured by CSA had limited use as CSA had no means of manufacturing the ammo.
Henry breech-loading rifle*	20	1,800	600	350	Faster than Spencer but more likely to jam.

*Omitted from firing tables for simplification.

Infantry fire-fight for possession of the Confederate farm. Artillery is being advanced to support the Federal infantry, who have the disadvantage of being in the open.

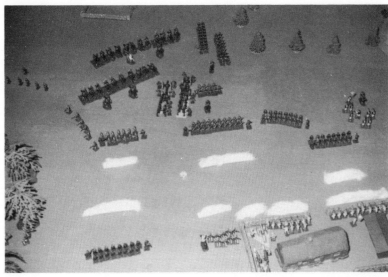

were armed mainly with muzzle-loading smooth-bore muskets, but by the autumn of 1862 the Union Army was equipped with muzzle-loading rifles and the CSA managed to equip its infantry with rifles at approximately the same rate, either by capture, import or manufacture. Production of breech-loading rifles for the US Army was begun in 1862 but their distribution was extremely limited in 1863 and even in May 1864, at The Wilderness, Grant had only 11 regiments armed with the Spencer repeating rifle—about three per cent of his army. Not until late 1864 should wargame Union armies have repeaters, and then only one or two regiments should be armed with this weapon.

It may be assumed that in 1861 the Confederacy's infantry weapons were inferior to those of the Union troops; perhaps 85 per cent smooth-bores to 15 per cent rifles, against the Union's 75-25 per cent. In 1862 there should be approximate equality with 50 per cent rifles, and in 1863 with 100 per cent rifles. The Confederacy's infantry weapons would be inferior again in 1864-5 as the US Army began receiving repeating rifles in larger numbers. Elite regiments tend to receive the best weapons, Springfields and Enfields always remain the most numerous

weapons, and Zouave regiments would have the Remington.

Cavalry firing

Cavalry was armed with sabre and revolver at the beginning of the war and at least two squadrons (or companies) in each regiment also had muzzle-loading carbines or rifles. By early 1863 all cavalry was equipped with carbines and from 1864 there was a gradual switch to breech-loading repeaters in the Federal Army. CSA cavalry relied on the Enfield carbine with its greater range and accuracy. Cavalry are not allowed to fire carbines or rifles while mounted in wargames, but when dismounted they fire as infantry. The inferiority of the carbine to the infantry weapons shows in the cavalry weapons table and is reflected in cavalry firing tables.

Sabre fighting was mostly abandoned after the first two years of the war—earlier in the western theatre—though the US cavalry continued to carry their sabres. Most cavalry carried two revolvers for close quarters fighting, the Confederates often having four or two and a double-barrelled shotgun. The latter would only stop a man at about 15 yards.

The revolvers were mostly Colts, Remingtons and Starrs, all of which

Cavalry weapons

Weapon	Rounds per minute	Ranges in yards		Remarks
		Effect-tive	Battle	
Enfield muzzle-loading rifled carbine	2-3	500	300	Widely used by both sides early in war. Preferred by CSA even late in war because of accuracy and rugged reliability.
Spencer breech-loading rifled carbine	10	450	300	Best cavalry arm of the war. Decisive in many cavalry actions.
Sharps M1859 & M1863 breech-loading rifled carbine*	8	500	300	2nd only to Spencer.
Burnside breech-loading rifled carbine*	about 8	450	250	3rd most popular.

*Omitted from firing tables for simplication.

Boys, this will be short but desperate

On the opposite flank the Confederate attack has closed up to the Federal farm and has broken the regiment to the right of the farm (seen routing). Federal cavalry has charged in support and are seen just after firing and before the mêlée. Note gaps created by cavalry firing.

Next phase of the cavalry fight: in the background some Confederates have ridden through the Federals, been hit by infantry fire, and are now charging back into the mêlée. In the foreground the second half of the CSA regiment has advanced to engage those Federals who have penetrated the first line.

held five rounds (six if placing one under the hammer) and had a rate of fire of 15 rounds a minute. However, reloading in the saddle during a mêlée was impracticable and therefore we only allow the cavalry to fire for the number of moves corresponding to the number of revolvers carried. Thus the US cavalry generally had a sabre and two revolvers and can fire for two moves and must then spend a move reloading. CSA cavalry with two revolvers and a shotgun, or four revolvers, may fire for four moves and then spend two moves reloading.

The maximum range of the revolvers was 300 yards, but effective range was only 50 yards with a 25 yard accurate battle range. As the revolvers will only be used in a charge or mêlée, firing is allowed only at the 25 mm range, ie just before the two sides meet. This applies to the shotgun also.

One other cavalry weapon should be mentioned — the lance! This was carried by the 6th Pennsylvania (Rush's Lancers) until 1863 and possibly by the 26th Texas and other units named as lancers. However, it was never used in any large combat and it is not realistic to include lancers in your army, no matter how dramatic they may look: it was not their kind of war!

Indians firing

The Indians of both sides carried a wide range of weapons: bows, tomahawks, lances or spears, and sometimes issue sabres, revolvers and muskets. The latter were of poor quality. CSA Indian cavalry also liked the shotgun.

Indians rarely opened fire at long ranges in the pre- and post-war Indian wars, and 100-150 yards was a normal range for them to commence firing. It is best therefore to arm all Indian infantry with the early smooth-bore muzzle-loaders. Dismounted cavalry also have this firearm. For mounted firing,

Maximum effective ranges*

Artillery piece	Shot	Shell Min	Max	Canister	Remarks
6pdr Napoleon smooth-bore	1,000	250	800	250	Used mainly by CSA in 1861. Replaced as soon as possible by 12pdr or 3-inch rifle.
12pdr Napoleon smooth-bore	1,200	250	900	300	Most popular smooth-bore. Reliable and effective.
12pdr Whitworth rifle	2,000	—	—	300	Used by CSA. Exceptionally accurate.
3" Rodman rifle	1,800	250	1,400	300	From 1863. Popular with Union Army, favourite of their light batteries.
20pdr Parrott rifle	1,900	250	1,400	300	The basic piece, used by both sides.
30pdr Parrott rifle	2,200	250	1,600	300	Used by heavy (reserve) batteries.
12pdr smooth-bore howitzer	1,070	150	750	250	
24pdr howitzer	1,300	300	800	300	
32pdr howitzer	1,500	500	900	300	

*All at approximately 5 degree elevation.

Boys, this will be short but desperate

The Parrott rifle.

Indians have an edged weapon and one revolver, so may only fire one move and must then spend half a move reloading. However, an allowance for skill with the edged weapon makes them as good as US cavalry veterans. For bows see rules summary.

Artillery firing

The main artillery pieces used from 1862 were: **US Army** 3'' Rodman and 20pdr Parrott rifles, and 12pdr Napoleon smooth-bore; **CSA** 12pdr Napoleon smooth-bore and 12pdr Whitworth rifle in equal proportions, and the 20pdr Parrott rifle. By 1863 both armies usually had between a third and half of their pieces 3-inch or 20pdr rifles, the remainder smooth-bores. Heavier batteries can be represented by the 30pdr Parrott for both sides if required.

It is a matter of choice whether to include howitzer batteries in your army. The scaling down does not permit many batteries and in a small army you will find the howitzer battery is never where you need it, whilst you could

Federal artillery firing ball against the advance on their left flank prior to the cavalry mêlée. The shot is on target (5 or 6 on deviation device), and the player has guessed the correct range, so the shot will take two men from the centre of the regiment.

American Civil War Wargaming

desperately use another smooth-bore one in its place!

Both rifles and smooth-bores could fire shot, shell and canister. (The term shell is used here for the projectile which fragmented *and* for spherical case, which scattered musket balls.) The shot relied on its speed and weight for cutting down men, and was devastating amongst closely packed troops. It did not hit the ground and stop dead, but rolled or bounced for some distance. The rules use this roll for assessing casualties. The player estimates the range to his target, measures that distance (and no more), then takes all figures within the roll length along the line of fire, the latter being determined by a deviation device. Whether a hit or miss is scored depends largely on the player's skill at estimating ranges, but once a shot has been fired at a stationary target the range is known. Units to the rear of the target may also be hit by the rolling cannonball. Each gun may fire one shot per move. Because muddy ground prevented the ball from bouncing and reduced the roll, the roll length is reduced by half if the ground is muddy.

Solid shot was ineffective against

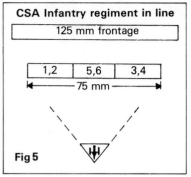

Artillery line of fire device. A dice throw indicates whether the shot goes on target (5 or 6), to the left (1 or 2) or to the right (3 or 4.) The roll of the cannonball takes effect along this line from the range estimated by the player.

entrenched infantry from the range necessary to protect gunners from the infantry's rifles, and therefore shell came to be used more and more in the second half of the war. It was also used as long-range canister against infantry attacks, and statistics show that in fact spherical case was no more effective

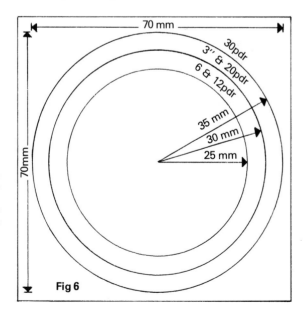

Scatter device for shell / spherical case, using transparent material such as Perspex. 50 per cent of the figures within the appropriate circle are removed; dice score of 4, 5 or 6 to save officers.

Federal advance on the CSA left flank receiving shell. Range correct, deviation of shot to left and scatter device indicates casualties.

than shot *unless* used for such plunging fire. Assuming shell is always fired with a high trajectory in wargames, we need a device to show the scatter of the projectile. Ranging and deviation of line of fire are as shot, and again all guns fire once per move.

Another device is needed to show the scatter of canister, in this case tapering off after the half-way mark because of the lessening effect of canister after that point. Canister could be fired twice as fast as other projectiles by the smooth-bores, so half the figures within the device are removed if it is a rifled piece firing (throw 4, 5 or 6 on a dice to save officers) but all the figures for smooth-bores.

The range and type of ammunition

being fired must be written in the orders before firing commences. If it is not, then the range and ammunition used for the last firing move must be used. If the gun has not fired before, then it loses its firing that move.

A word about comparative performance of rifles and smooth-bores. The rifle could out-range the smooth-bore, hit harder and was more accurate, but there were problems with malfunctioning of ammunition or fouling of the rifling, which caused loss of accuracy. When firing shell, rifles also tended to drive the projectiles so deep into the ground that their burst was ineffective. At Bull Run General Imboden referred to his area as looking 'as though a drove of hogs had been rooting for potatoes.' These faults to a large extent neutralised the technical superiority of rifled guns, although they retained the advantage of longer range.

Smooth-bores were much less accurate, had a shorter range, but were rugged and reliable, and were murderous weapons at close range. Imboden claimed one battery of smooth-bores was worth two of rifles and many gunners agreed with him. Certainly at close range the smooth-bore was supreme.

For these reasons no advantage has been given to rifled pieces in the rules beyond their longer range, and this is perhaps balanced by the effect of smooth-bores at close range.

Batteries may only fire within the angle of their template, and if wishing to change front take a whole move to do so. This is because the guns need to be limbered up, the battery wheeled to the new angle—with the outer gun having to move a long distance, as when any formation wheels, then unlimbered again.

Canister device. The player firing is allowed to position the device to his best advantage.

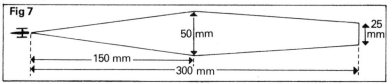

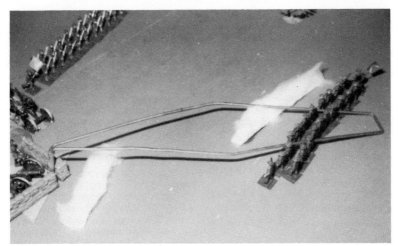

CSA infantry arrive within range of canister. Eight figures are within the device, so four are removed, as it is a Federal rifled artillery piece firing. CSA smooth-bores would take all eight figures.

Counter-battery firing

A hit with shot may damage a gun or limber but as the models represent a battery it is not possible to 'destroy' a model gun or limber with one hit. However, the battery's firepower must be affected and therefore the casualties inflicted by that battery must be reduced: in a four-gun battery the loss of one gun would mean ¾ casualties taken, loss of two guns ½ casualties taken, etc. In a six-gun battery the

On the CSA left the Federal artillery has come into action and silenced the CSA battery at the farm to allow the infantry to press home their attack.

Boys, this will be short but desperate

same principle applies. Limbers destroyed mean the guns cannot be moved and it may become necessary, if moving a battery, to abandon a gun or guns for this reason. The same casualty percentage reduction applies to these batteries.

Hits by shell will kill crew members. Normal casualty rates apply, with the device, and it follows that if a four-gun battery loses a gunner (25 per cent casualties) it will no longer be able to man all guns efficiently and casualties inflicted by that battery should again be reduced by 25 per cent.

Indirect firing

Howitzers may fire 'blind' from behind cover but must have an observer of one figure capable of seeing the target and within 200 mm of the howitzer battery, ie one man every six yards in scale. The observer may serve more than one battery provided the batteries are side by side.

Machine-guns

Machine-guns were available to both sides during the Civil War but were seldom used due to problems of mobility and ammunition supply. The table shows known details of the various guns' performances and rules could be worked out from this data. I do not use machine-guns in my own games as they played a small part in the war and were limited to sieges, which do not transfer to the wargames table very well in this period.

Infantry mêlées

When infantry advance to engage in hand-to-hand fighting, both sides may fire at 100 mm (75 mm for smoothbores) as closing, those advancing losing 50 per cent of their total move to do so. Hand-to-hand combat then takes place between all figures in contact, ie base to base. Gaps in the lines are not closed up, but penetrated by the opponent's figures if wished to the full length of the move outstanding or until reaching another enemy figure. A dice is then thrown by each player for each pair of figures in contact, the highest scorer being the winner. Add one to the scores of veterans and all Indians, deduct one from that of

Machine-guns

Weapon	Ammo	Rounds per min	Range	First used in action	Remarks
Billinghurst Requa Battery gun	.58 cal	7 volleys of 25 shots	Over 1,000 yds	Built late 1861	Loose powder used and hazards of sparks & rain.
Agar machine-gun	.58 cal	120	Unknown Deadly at 800 yds	29/3/1862	US weapon; 2 taken by CSA June 1862. Unpopular because of unreliability.
Williams machine-gun	1 pound shell or canister	20	Up to 2,000 yds	31/5/1862	CSA weapon. 42 guns in 7 batteries. Very reliable but ammo problems restricted use.
Vandenberg volley gun	Various musket cal	—	90% hits at 100 yds	—	—
Gatling machine-gun	.58 cal	250	—		12 purchased by General Butler in 1864; Hancock also 12, Porter 1. Not very accurate. Improved M1865 too late for Civil War.

American Civil War Wargaming

On the CSA right the Federal cavalry is finished off. Federal infantry in the farm has been driven off by musketry but the artillery remains in action, though depleted.

recruits. Nothing happens if there is a draw. Winners advance into the spaces left by the removed losers.

If attacking unformed infantry, or infantry in the flank or from the rear, the dice scores of the attackers are doubled for the first move.

Cavalry mêlées

These are more confused but use the same principle of individual figure combats. First of all it is essential every unit has recorded its weapon potential: I find it easiest to standardise: sabre and two revolvers for Federal cavalry, four revolvers for Confederate cavalry, a revolver and edged weapon for Indians. Cavalry fired their weapons just before contact, at point-blank range, getting off a large number of shots but firing at speed from horseback. It is assumed 50 per cent of the firing will be effective under these conditions. Therefore, advance to within 25 mm of

the enemy, dice for each figure firing: 4, 5 or 6 removes the enemy figure directly ahead and within 25 mm.

As a result of this firing there may now be a gap, through which the victor continues; a double gap with both figures dead; or no gap at all. If there is a double gap, reserves may close up if desired. Where the victor rides through a gap, he continues to the end of his outstanding move, allowing a deduction for reining up to turn round and prepare to repeat the manoeuvre. If there is no gap the two figures still facing each other now use edged weapons (US and Indians) or clubbed weapons (CSA). Again dice, highest score wins, as in infantry mêlées. Where edged weapons are successful the loser is removed: where clubbed weapons are successful the loser is dismounted and may be taken prisoner if his unit loses the mêlée, or remount if it wins. Veterans and Indians add one, recruits deduct two as for infantry

Boys, this will be short but desperate

The CSA advance on the right has been successful but on the opposite flank the rebels are hard pressed. Now the Federals begin a counter-attack, aimed at isolating the CSA right. We must leave the game there, undecided, but with the numerically inferior Confederate forces under strong pressure.

mêlées. Again, the winner of the combat will ride through to rein up with the remainder of his unit. At this stage Indians are usually better off withdrawing, as their weapons are unloaded.

If the manoeuvre is repeated, at the end of the second round of mêlée US cavalry will have their weapons unloaded and be at a disadvantage to Confederate cavalry, who still have two moves with loaded weapons. This tends to cause cavalry mêlées to be restricted to one or two moves, as they should be, and it is always advisable for the US player to keep back one move of loaded weapons rather than risk being caught with weapons unloaded for a further two moves by the CSA cavalry. Again, this causes US cavalry tactics to be rather cautious, as they were until they heavily outnumbered the Confederate cavalry in the last year of the war.

Cavalry charging unformed infantry or gunners fire at 25 mm, then mêlée with sabre or clubbed weapon with the nearest figure, doubling their dice score for the first move.

After all mêlées, both infantry and cavalry, all units need a full move to rally and reform. Reloading may also be carried out during this move.

Morale

Possibly the most important part of any rules is Morale; how the troops react to various situations on the wargames table. The most common type of morale rules require constant consultation of charts and a lengthy process of addition and subtraction to arrive at a morale value. This method is very efficient (examples of variations on the method may be found in any rules or books on wargaming) but even veteran players have to consult the charts for every morale check and this does slow down the game.

Personally I prefer a much simpler system based on figure value and in practice, during countless games, I have found this to often have almost identical results to the complicated charts, and in some cases it has actually proved more realistic — mainly because it dispenses with the unpredictability of dice.

The basis of my system is that every

American Civil War Wargaming

Cavalry mêlée viewed from the Federal side. In the foreground are men of the CSA front rank who have killed their opponents, passed through the Federal line, and are reining round to repeat the attack. A single Federal figure has pierced the rebel line, only to run into infantry fire. In the background the CSA have again created a gap and eventually the Federal cavalry was utterly defeated.

officer figure has a points value of 4, every sergeant 3, every corporal 2, and every private 1. (This can be simplified by giving values only to officers and privates.) The result is a points or morale value as shown below. Here can instantly be seen the greater morale value of the CSA infantry regiments over all Federal infantry except Wisconsin regiments, which is as it should be.

US cavalry battalion—12 figs: 10 privates, 1 corporal, 1 sergeant + 2 officers = 23 points; US Sharpshooters—8 figs: 6 privates, 1 corporal, 1 sergeant + 1 officer = 15 points; US Regular regiments—16 figs: 14 privates, 1 corporal, 1 sergeant + 2 officers = 27 points; Wisconsin regiments—20 figs: 16 privates, 2 corporals, 2 sergeants + 2 officers = 34 points; US volunteers and militia—15 figs: 12 privates, 2 corporals, 1 sergeant, + 2 officers = 27 points; US Indian units—15 figs.*

CSA cavalry regiments—20 figs: 16 privates, 2 corporals, 2 sergeants + 2 officers = 34 points; CSA infantry regiments—20 figs: 16 privates, 2 corporals, 2 sergeants + 2 officers = 34 points; CSA Indian units—10 figs.*

No allowance is made in this system for Negro troops, as on all occasions studied their value was equal to that of white troops.

Recruits and experienced troops break and run at double quick time when their morale value reaches the number indicated in the morale table. Veterans also break but retire in good order at common time. The breaking values set out in this table are based on 33⅓ per cent losses for recruit units, 50 per cent losses for experienced units, and 66⅔ per cent losses for veterans. Although Indians are classed as veterans for mêlées, here they are classed as recruits, as they could not be expected to stand and endure a heavier percentage of casualites.

*Contrary to popular belief, Indians did not always run away the instant a chief was killed, but would fight until a certain proportion decided to quit, when the rest would follow. Therefore, the morale value takes no account of the rank of individuals.

Boys, this will be short but desperate

A depleted Federal Zouave regiment (in light blue greatcoats) breaks and runs from the firing line. A general attempts to rally them to plug the gap.

The percentages themselves are based roughly on known proportions of losses in battle. For example, in World War 2 losses of ten per cent were barely tolerable yet in the Civil War units frequently lost 50 per cent of their strength in one battle—not because their time in action was prolonged but simply because firepower had increased so much—and occasionally regiments lost in the region of 80 per cent. (At Antietam the 1st Texas lost 82.3 per cent; at Gettysburg the 1st Minnesota 82 per cent; at Bull Run the 21st Georgia 76 per cent; and again at Gettysburg the 141st Pennsylvania 75.7 per cent.)

Regiments do not necessarily lose the percentage of men shown above before breaking, as the loss of officers and NCOs has a drastic effect on morale, and therefore with this system it is possible to have a numerically strong regiment whose morale fails because of lack of leaders.

Artillery are considered to remain in action unless ordered to withdraw. In practice a player will almost always move his guns if they are suffering heavy casualties, so as to preserve

Morale table

Type of unit	Morale value	Value at which units break		
		Veterans	Experienced	Recruits
US cavalry battalions	23	8	11	15
US sharpshooters	15	5	8	10
US Regular infantry	27	9	14	18
Wisconsin regiments	34	12	17	22
US vols & militia	27	9	14	18
US Indian units	15	—	—	10
CSA cavalry regiments	34	12	17	22
CSA infantry regiments	34	12	17	22
CSA Indian units	10	—	—	7

what firepower they have left and if possible use them again from a more sheltered spot.

Retreat continues for all units until they leave the table, unless a general officer joins them, when a dice may be thrown. For CSA troops a 5 or 6 will rally recruits, 4, 5 or 6 rally experienced troops, and 3, 4, 5 or 6 rally veterans. Union troops require a 5 or 6 to rally recruits, 4, 5 or 6 to rally experienced troops or veterans. This reflects the better leadership of Confederate forces and gives them a slight edge to counter numerical inferiority.

All Indians may be rallied by a throw of 3, 4, 5 or 6 in any move, without the presence of a general. This counters their low breaking point and reflects their hit-and-run tactics.

If a general fails to rally a unit he may remain with it, attempting to rally each move until the unit leaves the table. However, the general is not obliged to leave the table with it. Likewise, a general does not have to remain with a unit he has rallied.

Those units which have been rallied act as normal unless coming under fire, when they must throw a 4, 5 or 6 to stay for every move under fire. If they fail to obtain this score, they begin to retreat again and fresh attempts must be made to rally them.

Because of the poor chances of holding broken troops, especially recruits, it will be found best to allow them to leave the table except in the case of regiments which are numerically strong but whose morale has failed due to loss of leaders. Here a general may decide to remain with the unit permanently, and this is often where general officer casualties occur. It will also be found under this system that it pays a player to advance fresh regiments through his depleted ones after an attack, and retire the latter to the second line where they will continue to play a significant part in the battle, rather than drive a regiment on until it is utterly shattered.

Officers and NCOs will become casualties in mêlées and as a result of artillery fire. Such casualties are haphazard. There is no provision for such casualties to be taken during infantry firing but if desired a dice can be thrown for each move that a unit under fire loses three men or more. A throw of six removes an officer or NCO. In practice few officer or NCO casualties will be taken this way and it is an unnecessary complication, for there is a sufficient proportion of officers and NCOs killed by artillery fire and mêlées to have their required effect.

Boys, this will be short but desperate

five

Rules summary

This chapter is designed to sum up what has gone before and provide a ready reference for playing wargames. It is not a complete summary of the rules and should be used in conjunction with the detail in chapter four.

Scale

1 model soldier = 25 real men.
1 model gun = 1 battery.
1 mm = 1 yard horizontal scale.
4 mm = 1 foot vertical scale.
1 game move = 2½ minutes real time.

Classification of units

Veterans All Union sharpshooters and Regular regiments, all artillery, and any regiments with a long record of 'active service.' **Experienced troops** Any troops with previous battle experience. **Recruits** Untried troops.

In the first half of the war 50 per cent of the troops should be experienced and on average 25 per cent recruits, 25 per cent veterans. However, it is best to give the smaller CSA army more veterans and fewer recruits. In the second half of the war the percentage of recruits should remain about the same, but veterans would increase at the expense of the experienced class.

Orders

Before a game commences the Army commanders should write orders outlining areas of operation and objectives for each of their divisions or brigades. Each division or brigade commander should write basic orders for all regiments under command. Once a game begins these orders may only be changed by the general commanding or an ADC carrying written orders. Regimental orders, written each game move on a separate sheet, should concern themselves only with local matters, ie whether to advance, fire, charge etc.

Setting up the table

Terrain, nature of obstacles, numbers of units involved and their classification, and number of general officers, should all be determined before play commences.

Sequence of play

(With simultaneous moves by both armies.) 1. Write orders.* 2. Declare charges. 3 Carry out moves. 4. Artillery firing. 5. Infantry and cavalry firing. 6. Mêlées. 7. Morale and prisoner decisions.

Weather

Dice, score of 6 = heavy rain, and roll of shot is halved. Dice again, 1, 2, 3 = no wind, 4, 5, 6 = wind.

'Fog of war'

If there is wind or heavy rain this dispels smoke, otherwise each time a unit fires place a piece of card or cotton wool, equal to the frontage of the unit, 100 mm in front of the unit firing. If firing the next move, place the smoke 70 mm from the unit: if firing a third consecutive time place the smoke 35 mm from the unit. The position of the smoke indicates how far the unit can see. If a unit misses a move of firing, advance the marker one stage. The system is rather complex and sometimes annoying in a fast game, but your regimental orders keep track of firing and the system does discourage indiscriminate firing at poor targets.

Markers

Where units are not visible to the enemy, ie in woods, buildings or behind high ground, markers of the same size

*In solo games alternate moves are employed: attacker moves first, defender moves, joint firing and mêlée for simultaneous casualties.

American Civil War Wargaming

as the ground occupied by the unit may be placed on the board. Spurious markers, to a ratio of one false to every three real, may be used to confuse the situation more.

Visibility

Restricted to 100 mm in woods. Therefore units in woods need not be placed on the table until identified by an enemy within 100 mm, and even then only the front rank need to be placed in position, the marker remaining until all the troops it represents have been placed on the table.

Troops on hills of more than one contour may see over woods and buildings but the first 75 mm from the far edge of the wood or buildings are dead ground.

Movement

Cavalry in line 300 mm; in column 375 mm; charge move 475 mm.

Infantry in line 150 mm; in column and charge move 275 mm.

Skirmishers in extended order 225 mm; retreating from attack 275 mm.

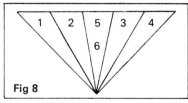

Fig 8

Charge deviation device, actual size.

Light batteries 375 mm; Field artillery 300 mm; Heavy batteries 250 mm; Wagons and siege artillery 150 mm.

Staff officers 480 mm.

Gunners on foot and infantry within fortifications move at 275 mm at all times.

Movement penalties

Deduct ⅓rd move to: cross a stream; cross a contour (all units); cross a hedge, fence, ditch or wall (cavalry and infantry only); to lie down or stand up (infantry and dismounted cavalry only); to unlimber a gun and come into action or to limber up.

Deduct ½ a move in woods.

'Smoke' in use during the opening phase of the wargame illustrated in the previous chapter. Here is seen a CSA feint on their left, prior to falling back to a strong defensive position at the farm while the main assault was launched on the opposite flank.

US forces under pressure at the end of a game. **Above** *The Federals' advance position across the river is about to be forced back (Move 14)* **Below** *CSA forces cross the river beyond the bridge (Move 16).* **Above right** *The final assault. (Note the batteries are advanced at each phase of the battle's climax.)*

All troops except skirmishers are unformed on leaving streams, buildings and woods and take ½ move to reform.

Where contours are less than 50 mm apart cavalry and artillery may not ascend or descend. Where they are less than 25 mm apart only skirmishers may ascend or descend.

Cavalry reining in after a charge and turning about deduct 75 mm.

Charging through smoke: advance the charging unit to the smoke, place

American Civil War Wargaming

deviation card by smoke at centre of charging unit. Dice and carry out remainder of move in direction indicated by the card. If missing target completely, carry on to end of move unless hitting another unit.

Cavalry charging over broken ground: classed as unformed and lose ten per cent every move, the ten per cent being dismounted and requiring a full move to remount.

Evasion: staff officers and skirmishers may move up to their maximum distance to avoid contact with the enemy. This is automatic and requires no written order.

Cavalry fighting on foot: deduct one third move to dismount or remount. One figure in four remains with horses, except in the case of Indians.

Changing direction: units change direction as ordered, the only restriction being no figure may move more than the maximum distance permitted for the formation of which it is a part.

Changing formation: units changing formation do so at their top speed. As the formations are company front and line, it is relatively simple for column to wheel into line and line to wheel into column. Changing formation therefore takes only half a move.

Formation definitions

All columns are in close order—two abreast and ranks close together. Troops in line are considered to be in open order, one or two ranks deep. Skirmishers are in extended order with a minimum of 25 mm between figures. Indians should always move as skirmishers.

Artillery movement

To discourage too much manoeuvring of artillery, any gun having part of its move left on reaching a new position loses that part of the move and may not fire until the next move. Gunners may move away from their guns but may not return to them and fire in the same move.

Staff

If a message is delivered, its contents may not affect movement, etc, until the following move.

Unformed

Troops are unformed if they are routed or rallying, or thrown into disorder by a manoeuvre, such as charging over broken ground or leaving woods, etc.

Firing

Infantry and dismounted cavalry may

only fire at targets within an angle of 45 degrees to their front, and guns at targets within the angle of their base plates. If wishing to fire at targets beyond these angles it is necessary to change front and lose a proportion of the firing. In the case of guns, no change of front *and* firing is allowed in the same move.

Firing restrictions

No firing within 25 mm of friendly troops.

No firing through your own skirmishers.

Guns may not fire canister over the heads of friendly troops.

Cavalry may not fire carbines or rifles when mounted; Indians armed with bows may fire from horseback.

Troops lying down and armed with muzzle-loaders may only fire every other move.

Crossing targets

A unit crossing the front of another will receive 50 per cent of that unit's firing at the shortest range, 50 per cent at the longest range.

Method of infantry and cavalry firing

Once per move, unless otherwise stated, removing casualties as shown by the firing tables.

US muzzle-loading smooth-bore

Range	Veterans			Experienced			Recruits		
in mm	Firing	Dead	%	Firing	Dead	%	Firing	Dead	%
75	2	1	50	5	2	40	3	1	33 1/3
100	5	2	40	3	1	33 1/3	4	1	25
150	3	1	33 1/3	4	1	25	5	1	20
200	4	1	25	5	1	20	7	1	15
250	5	1	20	7	1	15	10	1	10
300	7	1	15	10	1	10	15	1	7 1/2

Springfield

Range	Veterans			Experienced			Recruits		
in mm	Firing	Dead	%	Firing	Dead	%	Firing	Dead	%
100	5	2	40	3	1	33 1/3	4	1	25
150	3	1	33 1/3	4	1	25	5	1	20
250	4	1	25	5	1	20	7	1	15
350	5	1	20	7	1	15	10	1	10
500	7	1	15	10	1	10	15	1	7 1/2
750	10	1	10	15	1	7 1/2	20	1	5
1000	15	1	7 1/2	20	1	5	—	—	—

Remington

Range	Veterans			Experienced			Recruits		
in mm	Firing	Dead	%	Firing	Dead	%	Firing	Dead	%
100	2	1	50	5	2	40	3	1	33 1/3
150	5	2	40	3	1	33 1/3	4	1	25
250	3	1	33 1/3	4	1	25	5	1	20
350	4	1	25	5	1	20	7	1	15
500	5	1	20	7	1	15	10	1	10
750	7	1	15	10	1	10	15	1	7 1/2
1000	10	1	10	15	1	7 1/2	20	1	5
1200	15	1	7 1/2	20	1	5	—	—	—

American Civil War Wargaming

Spencer rifle

Range in mm	Veterans Firing	Dead	%	Experienced Firing	Dead	%	Recruits
100	5	3	60	2	1	50	
150	2	1	50	5	2	40	
350	5	2	40	3	1	33⅓	Not issued to
500	3	1	33⅓	4	1	25	recruits
750	4	1	25	5	1	20	
1000	5	1	20	7	1	15	
1500	7	1	15	10	1	10	

Because the Spencer had such a higher rate of fire, the percentage of casualties is increased to represent part of that higher rate of fire; for the rest, units armed with Spencers may fire twice per game move.

Sharps rifle

Range	Firing	Dead	%
150	5	3	60
250	2	1	50
350	5	2	40
500	3	1	33⅓
750	4	1	25
1000	5	1	20
1500	7	1	15
1800	10	1	10

Indian short bows

Range	Firing	Dead	%
50	2	1	50
75	5	2	40
100	3	1	33⅓
150	4	1	25
200	5	1	20
250	7	1	15

US Sharpshooters only: may fire twice per move to represent greater rate of fire.

Enfield carbine

Range in mm	Veterans Firing	Dead	%	Experienced Firing	Dead	%	Recruits Firing	Dead	%
100	5	2	40	3	1	33⅓	4	1	25
200	3	1	33⅓	4	1	25	5	1	20
300	4	1	25	5	1	20	7	1	15
400	5	1	20	7	1	15	10	1	10
500	7	1	15	10	1	10	15	1	7½

Spencer carbine

Range in mm	Veterans Firing	Dead	%	Experienced Firing	Dead	%	Recruits
100	5	3	60	2	1	50	
200	2	1	50	5	2	40	Not issued to
300	5	2	40	3	1	33⅓	recruits
400	3	1	33⅓	4	1	25	
450	4	1	25	5	1	20	

As with the Spencer rifle, the faster rate of fire is represented partly by the higher rate of casualties, partly by being able to fire twice per game move.

Artillery firing

| Artillery piece | Maximum effective ranges in mm | | | |
	Shot	Roll	Spherical case/shell	Canister
6pdr Napoleon	950	50	250 — 800	250*
12pdr Napoleon	1,200	75	250 — 900	300
12pdr Whitworth	2,500	100	— —	300
3-inch Rodman	1,800	125	250 — 1,400	300
20pdr Parrott	1,900	125	250 — 1,400	300
30pdr Parrott	2,200	150	250 — 1,600	300
12pdr howitzer	1,100	— †	150 — 750	250*
24pdr howitzer	1,300	— †	300 — 800	300
32pdr howitzer	1,500	— †	500 — 900	300

*Do not include last 50 mm of canister device when taking casualties.
†The plunging fire with the howitzer firing ball would be used primarily against fortifications.

The percentage column in these tables has been added for information only and if you decide to transfer the tables to postcards for easy reference the percentages can be left off.

Troops may wish to fire *and* move, in which case the number of casualties will be reduced. For example, five veterans firing at 150 yards with Remingtons kill two men. If taking half a move and firing, they would kill only one man. Troops armed with weapons which may be fired twice per move, may move and fire once.

Cover and casualties

Stone buildings, stone walls, entrench- ments and revetted earthworks are classed as hard cover. If troops behind hard cover are firing, they suffer casualties at one third rate. If they are not firing they do not suffer casualties.

Hedges, woods, fencing and wooden buildings are classed as soft cover and troops sheltering behind these, whether firing or not, suffer casualties at half rate.

Troops lying down or in extended order suffer casualties at third rate. Troops in open order suffer casualties at half rate.

Gunners and engineers at work are considered in open order, in close order when on the move.

CSA infantry and artillery line a stone wall to gain maximum protection from Federal fire.

American Civil War Wargaming

Union infantry sheltering in a wood and behind fences (soft cover) but with guns in a revetted earth-work, classed as hard cover.

Destruction of cover

All forms of cover may be destroyed by artillery fire and are therefore given values: hedge, fence, wooden bridges and wooden buildings, 2 points per 25 mm (length of bridge, longest wall of house); stone walls, bridges and houses and earthworks are given 6 points per 25 mm. 12pdr smooth-bores score one point per hit. Rifles firing shell take 3 points per hit, 2 points per hit with ball.

Wooden buildings and bridges may also be set on fire by shell: if a hit is scored roll a dice and a score of 5 or 6 sets the building alight. Any building or bridge set on fire must be evacuated on the next move and may not be used again. Buildings having half their value taken by hits are considered demolished and do not suffer further damage. Only a quarter of the original number of occupants may remain within the rubble.

Mêlées

Individual figure combats, dice for each pair engaged. Add 1 for veterans and Indians, deduct 1 for recruits. Highest score wins. In the event of a draw nothing happens. If attacking from flank or rear the attacker doubles his dice score for the first move.

Mêlées may be discontinued by either side at any time but if one player wishes to continue when his opponent withdraws, he simply moves forward and the mêlée continues.

All units need a full move to rally and reform after a mêlée.

Morale

See chapter four and morale table.

Pursuit

Infantry or cavalry pursuing unformed infantry remove one figure for every pursuing figure able to make contact. When cavalry pursues cavalry, a running mêlée takes place with the pursuer adding one to his dice scores.

Prisoners

Apart from the men taken in cavalry mêlées, prisoners may also be taken in infantry mêlées or may surrender if routed and being pursued.

Pursued units may just declare their

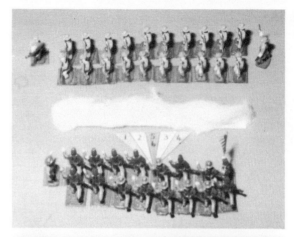

Union infantry about to mêlée with Confederates. The Confederates have been firing earlier and therefore the charge deviation device is used to align the Federals for the mêlée.

Wargames buildings detachable from their bases so that figures may be placed under them without losing their position on the table. The roof of the saloon is also detachable to allow figures to be placed in the upper floor.

A wagon train wagon under conversion for an observation balloon team. The balloon may be made from a real balloon with a thimble as the basket, supported by wire instead of a 'rope' from the 'windlass.'

surrender and must then be escorted to the rear by their pursuers. Infantry in mêlées may surrender whenever they wish, but *must* surrender if surrounded and their morale breaks, unless throwing a six when they may fight to the last man with no quarter given or taken. Units whose morale breaks in mêlée but are not surrounded must yield one third of their remaining strength as prisoners.

Prisoners must be escorted to the rear by one man to every five prisoners and held at the rear by one man to every ten prisoners.

Buildings

The number of men allowed in a building should be decided before play begins. Normally it is as many figures as will go into the ground space occupied — usually no more than a regiment. The figures can be represented by a marker.

Figures within buildings have their rate of casualties from firing reduced (see above) and in mêlées for possession of the building their casualties are halved. As casualties are deducted from the marker, and it is not mandatory to place the full quota of figures within a building, your opponent will often have no idea what effect his firing is having, or indeed if there is anyone left alive inside.

Engineers

Engineers move at 225 mm at all times. They may defend themselves by rifle fire or in mêlée but may not in any one move work as engineers *and* fight. If caught whilst working they must surrender.

They may be reinforced by infantry labourers at a rate of three infantry to one engineer. A unit of four may erect planking to repair a bridge in four moves; a river pontoon bridge for single line infantry in eight moves; a river pontoon bridge for single line cavalry or artillery in 16 moves; earthworks at the rate of 50 mm in four moves. The same unit may destroy by explosives a house, bridge or entrenchment after two full moves in contact with their objective: a dice throw is taken and if 1 scored there is a misfire and the unit must begin again. The same size unit may destroy 75 mm of hedge or 30 mm of entrenchments per move. All these actions require the engineers' wagon to be in attendance, ie within 50 mm of the work in progress.

six

Information

Sources

There is no one book which will give you all information you require on the American Civil War, but there is a wide range of books currently available on this popular subject and some of these supply information on uniforms as well as weapons, equipment and tactics. Two such books are Mike Blake's *American Civil War Infantry* and *American Civil War Cavalry*, both published by Almark. Blandford's *Uniform of the American Civil War in colour* by Philip Haythornthwaite also provides a great deal of information on weapons and the background to the war. R.F. Weigley's *History of the U.S. Army*, published by Collier-Macmillan in 1967, is still available and fills most of the gaps left by the books already mentioned.

Battles and Leaders of the Civil War, four volumes published by Thomas Yoseloff of New York, gives hundreds of accounts of the war but is expensive

Airfix Foreign Legion figures painted as 14th New York Volunteers with red kepi and trousers and white gaiters. The problem of the Zouave blouse is overcome by the fact they are wearing greatcoats, which are painted pale blue.

and not easy to borrow through the library system. It also gives some of the 'atmosphere' or 'feel' of the war, which assists in obtaining realism, or serves as inspiration when your eyes get tired of painting soldiers! Also in this category are Paul Jones' *The Irish Brigade* (New English Library, 1971), and *Soldier Life* edited by Philip van Doren Stern (Fawcett Publications, 1961.)

More specialist studies of the war, which are invaluable, are G.F.R. Henderson's *Stonewall Jackson* (Longmans, 1961) and *Braxton Bragg and Confederate Defeat* by Grady McWhiney (Columbia University Press, 1969.) These deal with the eastern and western theatres respectively.

Good summaries of the war are *The Civil War* by Harry Hansen (Mentor Books); *The Civil War* by J.S. Blay (Bonanza Books, New York); and *The Penguin Book of the American Civil War* by Bruce Catton.

Wargames rules

Most published rules do not separate the Civil War from the Napoleonic Wars sufficiently and do not allow for the variety of infantry firearms, the fact that cavalry no longer charged infantry on the battlefield, the differences in organisation, allow too high a percentage of repeating firearms, etc. Two sets of rules which I have tried and can recommend as worthy of experiment as a contrast to my own personal rules are: *American Civil War Rules* by the Confederate High Command, published and sold by Skytrex Ltd of Loughborough, Leics; and *Rules for the American Civil War c.1863* by the British Model Soldier Society Wargames Section.

Wargames clubs

If you do not have an opponent, and do not wish to play all your games solo, get in touch with your nearest wargames club: a club directory is published by Model & Allied Publications Ltd of Hemel Hempstead, Herts. Such wargames clubs have many

Badly needed limbers can be quickly made from the Airfix Wagon Train set front axle assembly, built up with balsa. Wagon train sets also provide several figure conversions.

advantages—for the beginner in particular they can provide a fund of information and cheerful assistance on problems by experienced players.

Model soldier manufacturers

Miniature Figurines Ltd, 28/32 Northam Road, Southampton. 25 mm metal figures. A complete range.

Tradition, 5a & 5b Shepherd Street, Mayfair, London W1. 25 mm metal figures. A complete range.

Rose Miniatures, 15 Llanover Road, London SE18 3ST. 25 mm metal figures. Infantry, cavalry and artillery.

Hinchliffe Models Ltd, Meltham, Huddersfield. 'Big' 25 mm figures. A complete range.

Hinton Hunt, Rowsley, River Road, Taplow, Bucks. 20 mm metal figures. A complete range.

K & L Company, PO Box 3781, Tulsa, Oklahoma, 74152, USA. A very full range of 20 mm metal figures.

C-in-C, distributed in UK by Micro-Mold Ltd, Station Road, East Preston, Sussex BN16 3AG. 20 mm metal figures. Limited range at present, but increasing.

Mounted Indians firing bows: made from Airfix Indians by pinning the upper halves of bowmen to the legs of mounted figures.

Building conversions are also possible and necessary, as there are few buildings on the market which are suitable for the ACW. Shown here is a jailhouse and courthouse building from the Airfix railway station, as modelled for an article in Airfix Magazine.

Airfix Products Ltd, available from Woolworths and most model shops. HO/OO plastic figures. Between 20 mm and 25 mm and not suitable for use with 25 mm metal figures, although they can be used with 20 mm metal figures if thicker card bases are used for the latter. A full range is produced which can be increased by conversions of the Cowboy, High Chaparral, Indians,

Wagon Train and Foreign Legion sets.

Most of these manufacturers advertise regularly in *Airfix Magazine, Military Modelling* and *Wargamer's Newsletter*. Up-to-date additions to their lists and the charges for their full catalogues may be found in these magazines. The recommended rules are also advertised in these magazines.

seven

The Peninsular Campaign

The Peninsular Campaign of 1862

There is not room in this book to go into the intricacies of wargame campaigning, where the earlier decisions of commanders and the results of previous battles play their full part in influencing subsequent events, but it is hoped this section might arouse sufficient interest to encourage readers to give campaigning a try; and once you have fought your first campaign it is most unlikely you will ever look back.

In the ACW period the Peninsular Campaign offers a good wargame campaign for beginners: it was fought in a relatively small area; there were a number of smaller actions as well as major battles; and it is well documented. It also includes Jeb Stuart's famous ride around the Union Army!

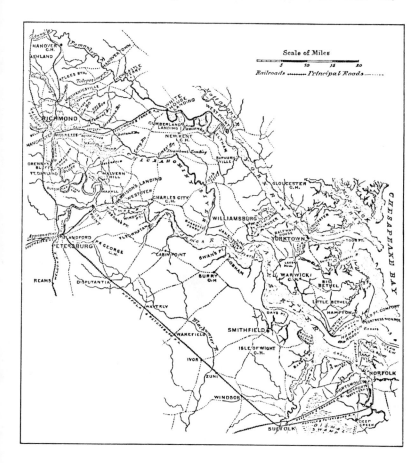

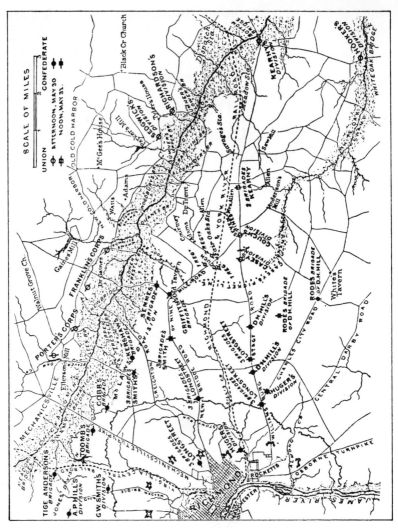

Seven Pines, May 31 — June 1

The Peninsular Campaign was launched by the Union in the spring of 1862 in an attempt to capture Richmond. During McClellan's build-up of forces in the peninsular the Union Army became divided into two parts by the Chickahominy River and on May 31 the Confederates launched an attack against McClellan's isolated left. Planned as a double envelopment, the attack degenerated into piecemeal frontal assaults due to poor staff work and after two days of fighting the battle ended in defeat for the rebels. General Johnston, severely wounded in the battle, was replaced by Lee on June 1. The map shows the preliminary positions.

Union artillery dump at Yorktown, McClellan's base for the Peninsular campaign.

American Civil War Wargaming

The Peninsular Campaign

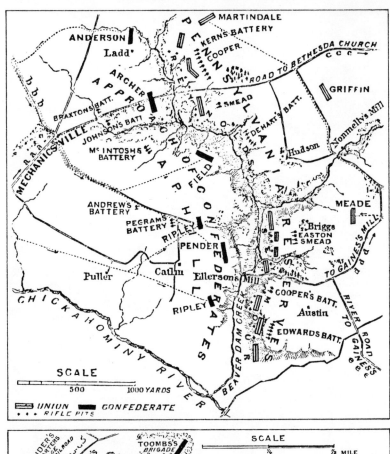

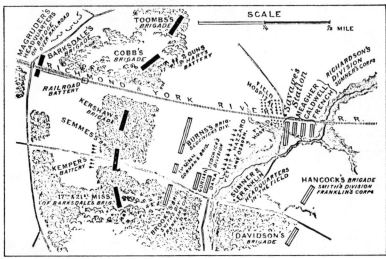

American Civil War Wargaming

Left Mechanicsville, June 26

On June 26 southern troops took the offensive against McClellan's forces and fought the battle of Mechanicsville, the first of a series of battles now known as the Seven Days' Battles. Again faulty staff work reduced the Confederate plan to piecemeal assaults and the Union troops under Porter could not be shifted from their strong position behind Beaver Dam Creek. However, learning of Stonewall Jackson's approach on his right, Porter withdrew to Gaines' Mill during the night and fought another successful defensive battle there on the 27th. **a** marks the approach of D.H. Hill and Longstreet from Richmond; **b** the approach of A.P. Hill; **c** the route of D.H. Hill to Old Cold Harbor on the 27th to attack the Union right; **d** the route of A.P. Hill to New Cold Harbor on the 27th to attack the Union centre; and **e** the route of Longstreet to Gaines' on the 27th to attack the Union left.

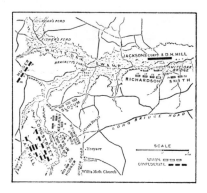

the James. The map also shows the dispositions of the troops during the artillery action at White Oak bridge at about 11am. The Confederates were again checked long enough for McClellan's trains to pass and the main body to take up a strong position on Malvern Hill.

Below left Savage's Station, June 29

On the night of the 27th Porter retreated across the Chickahominy and McClellan ordered a general retreat to the Union gunboats on the James River. To reach this position McClellan had to march south across White Oak Swamp, along narrow roads. His retreat was well planned and executed and the rearguard succeeded in holding off repeated Confederate attacks. The third of these, Savage's Station on the 29th, was inconclusive but the Union army had to destroy its stores before completing the crossing of White Oak Swamp.

Above right Frayser's Farm, June 30

This was the fourth clash, beginning about 3pm, fought by McClellan to protect his trains as they passed down the Long Bridge and Quaker roads to

Page 64 Malvern Hill, July 1

In McClellan's absence (he was at the new base on the James) Porter fought a brilliant defensive action in this, the greatest of the Seven Days' Battles. Within a short time the well-placed Federal batteries had silenced every Confederate battery within range and Lee's assault broke down into a series of unco-ordinated charges which were beaten back with heavy loss; about 5,000 Confederates died on the slopes of the hill.

Malvern Hill was a Union victory but McClellan ordered the retreat to the James to be completed. Union losses in the Seven Days' Battles totalled 15,849; CSA losses 20,614. Lee does not appear to have lived up to the reputation he was to be given; for, with the exception of Gaines' Mill, all Confederate attacks were repulsed; and yet at the end of the day McClellan's army had been forced back from Richmond by Lee's daring, and was now digging in on the banks of the James as if defeated, with McClellan asking Washington for a further 100,000 troops!

The Peninsular Campaign

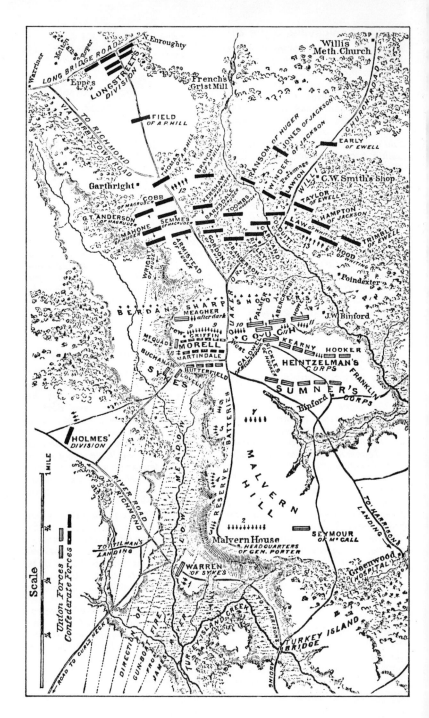

American Civil War Wargaming